Ana Mercedes Díaz de Iparraguirre

Climate change

Ana Mercedes Díaz de Iparraguirre

Climate change

environmental sustainability actions

Imprint

Any brand names and product names mentioned in this book are subject to trademark, brand or patent protection and are trademarks or registered trademarks of their respective holders. The use of brand names, product names, common names, trade names, product descriptions etc. even without a particular marking in this work is in no way to be construed to mean that such names may be regarded as unrestricted in respect of trademark and brand protection legislation and could thus be used by anyone.

Cover image: www.ingimage.com

This book is a translation from the original published under ISBN 978-613-9-41032-3.

Publisher:
Sciencia Scripts
is a trademark of
Dodo Books Indian Ocean Ltd. and OmniScriptum S.R.L publishing group

120 High Road, East Finchley, London, N2 9ED, United Kingdom
Str. Armeneasca 28/1, office 1, Chisinau MD-2012, Republic of Moldova, Europe
Printed at: see last page
ISBN: 978-620-8-03498-6

DEDICATION

To Almighty God and the Virgin Mary

For giving me the strength, tranquility, peace and encouragement to follow every path I have taken and for giving me encouragement through prayer to overcome the most difficult moments.

To my husband Miguel Angel

That from heaven, his love and his spirit accompany me on every road I travel to overcome the hard moments of his departure.

To my children

To Walter, whose love accompanies me from heaven, Annather for his show of affection and love. To my children Ana Carolina and Miguel Angel, who with their love, affection, understanding, company and material support, have been my motivation to complete the paths I have taken and to undertake new projects.

To my grandchildren

Katherin, Marbel, María del Rosario, Anthony, Santiago, Tomás, Mariangela and Sarita for the happiness and love they have given me.

To my parents and brothers

Mom and Dad who protect me from heaven, Bertha, Ramon, Eligia, Yesenia, Roger, Elvis, Mery and Erick for their understanding and the moments of pain shared and my pet Toby, for his faithful and loving company.

Ana Mercedes Díaz de Iparraguirre
E-mail: anamer49@yahoo.com
Mobile: + (58) 424 3177722
ORCID: 0000-0002-2241-81

INDEX

II

Summary

According to recent World Bank studies, climate change could cause the displacement of 216 million people within their respective countries by 2050; and that climate change could reduce crop yields, generating food insecurity; the agricultural sector being considered as fundamental to address climate change. The COP28 climate summit in Dubai concluded with a historic agreement that marks a significant step forward in the fight against climate change. The agreement, reached after days of negotiations, calls on all nations to abandon fossil fuels, measures aimed at limiting temperature rise to 1.5°C above pre-industrial levels. EuroNews (2023)

On the other hand, the agreement has been praised as a breakthrough, environmental groups and island islands have expressed their concern that there will not be a fair and rapid transition away from fossil fuels. These actions are aimed at caring for the environment, through investment in renewable and/or green energy, saving water, avoiding its waste, sustainable mobility and innovation in construction and architecture, giving greater impetus to sustainable products and technologies. In view of this, many innovators and companies are taking steps to build a greener future on the planet, through the advancement of the energy transition to boost the sustainable economy, sustainable innovations which explore the challenges of energy storage, Artificial Intelligence projects to address food waste, towards a more sustainable way of life. In that sense, scientists and scholars of the subject seek to raise individual awareness to change habits to stop the irreparable effects of climate change on the life of the planet. The work will be approached through the review of documents related to the researched topic.

Keywords: Climate change, sustainability actions, awareness raising

III

Introduction

Given that climate change is caused by greenhouse gases, which in turn is caused by the burning of fossil fuels for electricity generation, transportation, heating, industry and construction, the exploitation of livestock, agriculture, wastewater treatment and landfills, as well as the exponential growth of the population, which requires more resources that accelerate the increase of greenhouse gases in all production processes, has caused the planet to enter what the scientific community has called the Anthropocene: As a new geological era motivated by the impact of human beings on the Earth.

On the other hand, the destruction of terrestrial ecosystems and deforestation has caused forests and rainforests to rapidly disappear. These forests are natural carbon sinks that, through photosynthesis, absorb CO_2 and return oxygen to the atmosphere. Similarly, in marine ecosystems, the destruction of the oceans, which are also natural sinks for the production of oxygen, has accelerated the limit for the absorption of the amount of CO2 acidifying and thus causing the deaths and diseases of marine flora and fauna. On the other hand, the global increase in temperature has endangered the survival of the Earth's flora and fauna, including human beings.

It should be noted that climate change increases the occurrence of more violent weather phenomena such as hurricanes, cyclones, typhoons, droughts, rains, snowfalls, fires, death of animal and plant species, overflowing of rivers and lakes, appearance of climate refugees and destruction of livelihoods and economic resources, especially in developing countries; The heat also causes ice to melt at the poles, raising sea levels and threatening to submerge coastal coasts and small island states under water, causing more deaths, victims, displaced persons and material damage.

On the other hand, massive migrations have generated the figure of the climate refugee, not yet recognized by the United Nations, which is a reality that is estimated to reach one billion people by 2050. In view of this, a series of actions have been taken in order to limit the emission of greenhouse gases to avoid the increase in temperature and to be able to adapt to the environment through policies designed by international agreements. However, global warming generates consequences in physical, biological and human systems, causing variations in the climate that would not occur naturally.

The Earth has already warmed and cooled naturally on other occasions, but these cycles have always been much slower, while now, as a consequence of human activity, it has reached levels that in other times brought about extinctions in two hundred years. It is important to clarify that the scientific community, the UN and many international agencies have warned of the growing seriousness of the impacts of climate change and the need to act more decisively against its causes and consequences, especially in the rich and industrialized countries that have caused this situation the most.

On the other hand, according to UN (2023), the COP 28 summit held in Dubai between November 30 and December 12, 2023, closed the reflection process, called "Global Stocktake", on the climate action of the Paris Agreement in 2015. All this, in order to recognize the delays and failures of climate action to date, as well as the need and opportunity to act much more quickly and effectively to slow the advance of climate change and maximize resilience to its impacts, to try that industrialized countries accelerate the rapid and progressive abandonment of the use of fossil fuels, the burning of which is the main cause of the climate crisis, which increasingly threatens human life and ecosystems, habitats and species that we need to survive.

In view of this, they agreed to triple renewable energy capacity by 2030, while respecting biodiversity and the local communities around them in their installations:

- Absorb excess CO_2 from the atmosphere.
- Ensure security of supply of water, food and other basic human needs.
- Protect communities and sectors vulnerable to extreme weather events and to the generalized rise in global temperature and the increase in ocean levels and acidity.

In that order, at COP28 SEO/Birdlife, together with other movements of science and civil society, will ensure climate commitments to integrate the conservation and recovery of the natural environment for the coming years. Given that it will have catastrophic consequences for billions of people and ecosystems, as pointed out by the Intergovernmental Panel on Climate Change IPPC) (2023). As the world's climate has changed historically; global temperatures are rising at an unprecedented rate, with the last eight years being the warmest on record.

In view of the fact that the heat has increased the frequency and severity of extreme weather phenomena: more heat, drought and fires, more frequent extreme rains, melting processes that have caused the sea level to rise, it is necessary to continue fighting against climate change, generating contributions from all areas for environmental sustainability. The work through literary review seeks to reflect on climate change, environmental sustainability actions to raise awareness on the subject.

Literature Review

I.- Climate change versus human rights

Everyone has the right to live in a clean, healthy and sustainable environment. As the climate crisis intensifies, this and other rights are increasingly threatened. Climate change exacerbates droughts, crop damage and leads to food shortages and higher food prices, and, after decades of steady decline, world hunger is on the rise again. This scarcity increases competition for resources and can lead to population displacement, migration and conflict, which in turn lead to other human rights harms.

Already vulnerable, less fossil fuel-intensive communities such as subsistence farmers, indigenous peoples, and those living in low-lying island states facing rising sea levels and more intense storms often suffer the most from the consequences of climate change and are most often threatened in their right to health, life, food and education. Global warming affects many other rights in countries at all income levels, for example, with significantly worsening air pollution.

This also means that disease-transmitting mosquitoes are spreading to new areas. In that order, extreme heat causes deaths of people working outdoors, and increases mortality rates in residences and medical centers. Relatedly, in high-income countries, the damage caused by fossil fuel extraction and climate change often occurs in so-called "sacrifice zones," where communities that are already discriminated against often suffer harmful pollution; and disinvestment means that public infrastructure is ill-equipped to withstand extreme weather events.

However, with regard to human rights in the United Arab Emirates, being a major fossil fuel producer, it has a dismal human rights record which poses a threat to the success of the summit.

On the other hand, the promise to allow "different voices to be heard" at COP28 is inadequate and serves to highlight the UAE's normally restrictive human rights context and the severe limitations the country places on the rights to freedom of expression and peaceful assembly.Ohchr (2023) Concerns about the closure of civic space, and the possibility of digital spying and surveillance. Amnesty International has prepared a comprehensive report on the human rights situation in the UAE.

The COP must be a forum in which the right to freedom of expression and peaceful demonstration is respected and civil society, indigenous peoples and communities and groups affected by climate change can participate openly and without fear. In light of this, Emirati citizens and individuals of any nationality must be able to freely criticize states, companies and policies, including those of the UAE, in order to help shape policies without intimidation. It should be noted that the UAE is one of the 10 largest oil producing states in the world and is opposed to the rapid phase-out of fossil fuels.

The fossil fuel sector generates enormous wealth for relatively few corporate actors and states, which have a vested interest in blocking a just transition to renewables, and in silencing those who oppose them. COP28 was chaired by Sultan al Jaber, who is also the CEO of ADNOC, the UAE's state-owned oil and gas company, which is expanding its fossil fuel production. Amnesty International urged Sultan Al Jaber to resign from his position at ADNOC, as it believes there is a clear conflict of interest that threatens the success of COP28.

This is symptomatic of the growing influence that the fossil fuel *lobby* has been able to exert on States and the COP.28. In view of

Therefore, an agreement at COP28 to phase out fossil fuels in a rapid, fair and financed manner is critical to protect human rights. Moreover, government and business leaders can and must do much more to halt the increasing development of fossil fuel resource production, which is incompatible with States' human rights obligations and with the goal of limiting global warming to below 1.5°C.

In that order, many countries are investing in the expansion of renewables, but much more is needed to achieve a just transition that gives access to renewables to everyone. However, public financing for renewables, making the polluter pay, and mandatory electrification are policy approaches that can generate measurable impacts on emissions. In addition, there are a number of ongoing court cases related to climate change and rights violations. Amnesty International is involved in some of the cases, and these demonstrate that there are legal avenues to hold states and companies accountable.

Related to this, climate change campaigns and activism have achieved significant victories, highlighting that grassroots pressure on governments and corporations to stop investing in fossil fuels can help bring about a turnaround. Given that, it is youth and minority communities who are suffering from climate change-related human rights violations that are at the forefront of these initiatives.

II.- Fighting climate change starts at school

According to María Soledad Liora Schwartz (2023), climate change is generating an increase in temperatures and in the frequency and intensity of

extreme weather events such as heat waves, droughts, floods, landslides and tropical storms at an unprecedented rate in Latin America and the Caribbean. These changes in climate are generating devastating socioeconomic impacts in the region. Education has a key role to play in supporting efforts to decarbonize and increase resilience to climate change, in order to lead change and concrete actions to achieve it.

In 2015, the Paris Agreement on Climate Change committed countries to implement strategies to decarbonize their economies and reduce greenhouse gas emissions with the goal of keeping the temperature increase below 1.5°C compared to pre-industrial levels. At the same time, the countries identified strategies to increase resilience to the consequences of climate change on people, communities and their economies.At first glance, these two issues seem unrelated, and when thinking about climate change, topics such as renewable energies, circular economy, sustainable agriculture and resilience to climate disasters come to mind.

However, the link between the two is clear, not only because schools can do more to reduce their carbon footprint, but also because climate change is a factor that threatens the continuity of learning. For example, in 2021 hurricanes and tropical storms Eta and Iota damaged or destroyed nearly 1,000 schools in Honduras and Guatemala. These weather events meant that almost 700 schools had to be used as shelters. Meanwhile, Hurricane Mathew, in 2016, damaged more than 300 schools in Haiti and caused more than 100,000 students to lose learning due to damage and the use of schools as shelters.

In these circumstances, and due to the low technological incorporation, educational systems have not been able to implement quality alternative teaching methods that would allow for the continuation of the foresight of the

service in these emergencies.

On the other hand, education plays three main roles in accompanying and adding value to the countries' agenda of decarbonization and resilience to climate change. The following are three roles of education in addressing climate change:

1. **Educating in green citizenship:**

Education develops ~~key~~ skills during school age to equip children and youth with the knowledge, values and capacity for action in favor of the environment, what we call green citizenship. During school age, young people also acquire skills that enable them to access and succeed in jobs related to the decarbonization of the economy, such as solar energy, electric transport, circular economy.

2. **Resilience to avoid discontinuing learning**:

Education systems must be resilient and able to continue to operate in the face of extreme weather events and minimize disruptions to learning. Key here is to have schools that remain standing in the face of high winds or that are located in places that do not experience frequent flooding. In addition, well-developed distance learning systems allow children and youth to continue studying during these emergencies until they can return to the classroom.

3. **Sustainable educational service:**

It is essential to implement climate sustainability practices in school infrastructure and the operation of educational services to contribute to decarbonization goals.These strategies include the construction of schools that minimize the use of energy or water, electric school transportation, school gardens to grow food for school canteens, among others. On the other hand, 12 actions to fight against climate change from the educational systems are pointed out:

a.- How can green citizenship be developed at school age?

1. Reforming national curricula and syllabi to incorporate the development of environmental literacy throughout the school year.

Environment, biodiversity and climate change; valuing and respecting nature, environment and biodiversity; and pro-environmental behaviors. This includes encouraging extracurricular programs that allow students to complement and contextualize climate change education.

Expanding the supply of technical-vocational and higher education programs that develop skills for green jobs, in coordination with the growth and decarbonization strategies of the countries, the productive sector and job training systems.

3. Training teachers to have the knowledge and skills to deliver climate change education with effective, project-based, problem-solving teaching practices that foster lifelong learning.

4. Developing and adapting green citizenship skills measurement tools to monitor student learning and inform climate change education policy.

b.- How can the resilience of education systems be increased?

5. Including in the design, construction and operation of schools strategies for resilience to major climate hazards. For example, when faced with the threat of drought, installing rainwater collection and treatment systems, or when faced with rising temperatures, guaranteeing natural cross ventilation or solar protection measures.

6. Having contingency plans in place to prepare the education system for distance education models to provide continuity of educational service during weather emergencies until return to the classroom is possible due to weather contingencies.

7. Increasing socioemotional support to students before, during, and after extreme weather events in complement with health sector efforts.

c.- How can climate sustainability be achieved in educational buildings and in the provision of educational services?

8. Incorporating climate sustainability strategies in the design, construction and use of school infrastructure. For example, using solar panels or LED lights to save energy, self-closing faucets in bathrooms to save water, or using building materials with low energy and environmental impact (local, recycled and/or produced with less energy use).

Expanding the use of technology and digital systems for educational management in order to reduce the transportation of people and the use of paper for educational procedures and resource management. Or promoting distance education for certain educational modalities (flexible secondary education modalities, teacher training, remote tutoring) that reduce the transportation of students and teachers and thus reduce GHG emissions.

10. Ensuring that electronic devices are certified for low energy consumption and that their packaging, recycling and final disposal are environmentally friendly.

11. Reducing the environmental impact of transportation to school through the use of public transportation and electric school transportation.

12. Reducing the environmental impact of school feeding programs by, for example, using local and sustainably grown produce, as well as promoting the use of fruits and vegetables from school gardens.

In that order, promoting climate action and resilience to address the climate emergency and meet the Sustainable Development Goals, promoting environmental stewardship and counteracting climate change. through the reduction of greenhouse gas emissions. In this respect, many countries do not have sufficient resources to correct the damage caused by global warming, nor to adapt to its consequences and protect the rights of the population. According to the 2015 **Paris Agreement**, higher-income states have an obligation to provide them with support.

In 2009, higher income states, historically the largest emitters of greenhouse gases, pledged US$100 billion per year by 2020 to assist "developing" countries with emissions reductions and climate adaptation. So far, they have failed to meet this funding commitment, although fulfilling all the promises made and increasing funding for adaptation and social protection programs are essential to protect rights.

For years, higher income states refused to pay for loss and damage caused by climate change in "developing" countries. However, at COP27 it was agreed to create a Loss and Damage Fund. At COP28 it was negotiated how the fund will be directed and managed. In view of this, higher income states, with their role as creditors and regulators, and through their influence on the World Bank, will provide debt relief and/or loans with less stringent conditions, in order to help accelerate a just transition to renewable energies worldwide.

III.- Challenges of climate change as an opportunity to build a fairer and more sustainable world, according to the COP 28 Agreements

In the COP28 agreements to limit global warming to 1.5ºC, it was established to reduce global greenhouse gas emissions by 43% by 2030 and 60% by 2035 in relation to 2019 levels, in order to achieve net zero carbon dioxide emissions by 2050. The agreement also states that the end of fossil fuels will begin. A start

which, as the international CEO of the UN Global Compact, Sanda Ojiambo (2023), points out, leaves "much more to be done". Especially to those who opposed the phase-out of fossil fuels in the COP28 text, telling them that the phase-out of fossil fuels is inevitable, whether they like it or not.

In that vein, Antònio Guterres, Secretary General of the United Nations, said: "Although in Dubai we did not turn the page on the fossil fuel era, this outcome is the beginning of the end. He noted: Although in Dubai we did not turn the page on the fossil fuel era, this outcome is the beginning of the end. On the other hand, Simon Stiell (2023), Executive Secretary of the United Nations Framework Convention on Climate Change, said that he will analyze and evaluate whether the priorities have been met and discover the role played by the UN Global Compact in Spain in the face of such a pressing threat as climate change:

Conclusions of COP28

The first global assessment of the Paris Agreement leaves no room for doubt: we are still a long way from limiting the temperature increase to 1.5°C from pre-industrial levels. A context highlights that developing countries are particularly vulnerable to the adverse effects of climate change and that viable, effective and low-cost mitigation options are needed in all sectors. To limit global warming to 1.5°C, the agreement states that global greenhouse gas emissions must be reduced by 43% by 2030 and 60% by 2035 relative to 2019 levels, and achieve net zero carbon dioxide emissions by 2050.

The actions of companies must be geared towards mitigation, climate finance, adaptation and restoration of biodiversity. A summit, which leaves three conclusions:

a.- Consensus building is fundamental and rarely easy, and should even be easy:

.   COP28 reached an unprecedented agreement on the Paris Agreement, although it does not agree to "phase out fossil fuels" completely and stops short of eliminating them. However, the direction is clear.

b.- The ambition of non-governmental actors, especially the private sector, is a sign, but needs more impetus.

There were many announcements at COP28 and it was clear that the private sector has a key role to play. Corporate actions around renewables and climate finance have a key role to play, but there is a need to Involve more business, have a more holistic approach to climate action, more collaboration with business to develop national climate action plans and have more ambition.

c.- Global and local efforts must ensure effective and equitable shifts to renewable energy.

However, no one should be left behind and this transition must be fair and equitable. A development in which companies play a crucial role.

The Dubai Agreement. Under the umbrella of COP28, 198 countries signed the Dubai Agreement.

A pact that recognizes the need for deep, rapid and sustained reductions in greenhouse gas emissions in line with 1.5ºC trajectories. In this way, the following agreements have been reached:

- Goal 2030: Triple the global renewable energy capacity and double the global average annual rate of improvement in energy efficiency.
- Coal reduction: Accelerate the phasing out of coal-based energy use.
- Zero emissions: Move toward net-zero emissions energy systems globally, using low- or zero-carbon fuels before or around mid-century.
- Fossil fuel phase-out: moving away from fossil fuels in energy systems in a fair and orderly manner, accelerating action on clean technology development: developing zero- and low-emission technologies such as renewable energy, nuclear power, and carbon capture and storage technologies.
- Current decade to achieve net zero emissions by 2050, especially in hard-to-reduce sectors
- Non-CO2 gas reduction: Substantially reduce non-carbon dioxide gas emissions globally, with a focus on reducing methane emissions by 2030.
- Sustainable transportation: Accelerate the reduction of emissions in road transportation through infrastructure development and the rapid adoption of zero- or low-emission vehicles.
- Elimination of inefficient subsidies: Eliminate inefficient fossil fuel subsidies that do not address energy poverty or just transitions.
- Doubling adaptation funding from 2019 levels by 2025, halting and reversing deforestation and forest degradation, establishing all parties to national adaptation plans by 2030,
- Reduce water scarcity, achieve resilient food and agricultural production,
- Increase resilience of infrastructure to climate change, reduce effects of climate change on poverty eradication, etc.

In terms of international cooperation, it recognizes the role of business and stresses the need to strengthen incentives, regulations and conditions to guide investments to achieve a global transition towards CO2 emission reductions. In addition, the Technology Mechanism will support development through training, knowledge sharing, technical assistance and the role of artificial intelligence for climate change.

On the other hand, the text establishes that the 198 countries committed to the Agreement will report their national contributions to address climate change every 5 years. Transparency and clarity in reporting is emphasized. Although the agreement does not impose penalties for non-compliance, it serves as a shared roadmap to address the urgency of changing energy and trade systems to avoid consequences of climate change.

3.- UN Global Compact Spain responds to the call for climate action

At COP 28 in Dubai, the UN Global Compact Spain stood out for its commitment and leadership in promoting climate action at the business level. One of the notable events organized by the UN Global Compact was Caring for Climate, a meeting of international executives focused on accelerating emissions reductions to avoid climate catastrophes. Sanda Ojiambo (2023), international CEO of the UN Global Compact, emphasized the need for global collaboration and innovative solutions. Other high-level participants included José Manuel Entrecanales, CEO of Acciona, Ignacio Sánchez Galán, CEO of Iberdrola, and Clara Arpa, President of the UN Global Compact Spain.

During the event, crucial issues such as the end of the war on nature and biodiversity and the energy transition to climate neutrality were addressed. Another highlight, SBTi's leading companies, looked at Spanish companies leading the way in science-based target setting. Clara Arpa, (2023) opened the event and highlighted collaboration as a transformative tool, highlighting the role of the Climate Ambition Accelerator program. Miguel Arroyo, head of the environmental area, held a dialogue with Ligia Ramos, regional head of SBTi in LATAM, on the exponential growth that the SBTi initiative has experienced in the last year.

The event closed with a panel moderated by Salome Zurabishvili, CEO of the UN Global Compact Georgia, on science-based goal setting, which was attended by Megan Morikawa, global director of sustainability at Iberostar Group, as well as Karen Tanaka, head of climate action at

AmBev. The Spain pavilion hosted the Corporate Leadership in Climate Action event, where business leaders gathered to discuss the essential role of business in addressing the climate crisis. Clara Arpa highlighted that more and more companies are making the fight against climate change a priority.

Spanish companies were urged to join the Forward Faster initiative. At this same event, Miguel Arroyo and Philippine Ménager, project manager for COP ambition at ECODES, shared the results of the Climate Yearbook 2023, analyzing the commitment of Spanish companies in climate initiatives. Of the total number of Spanish companies that have committed to SBTi net zero, 76% are partners of the UN Global Compact Spain. Likewise, 94% of Spanish companies included in the highest rating of CDP Climate and more than 200 companies included in the carbon footprint registry of the Spanish Climate Change Office.

To close the event, Lucas Ribeiro, global head of the *Climate Ambition Accelerator* program, moderated a panel discussion on milestones and supply chain participation in climate initiatives, with the participation of Yolanda Fernández, director of environment, sustainability, innovation and climate change at EDP, Mercedes Vázquez, head of climate change at REDEIA, Etienne Butruille, head of financial sustainability at Banco Santander, and Tracy Wyman, head of outreach and engagement at SBTi.

On the other hand, CEOE and UN Global Compact Spain organized a Meeting of Spanish Companies attended by the Third Vice-President and Minister for Ecological Transition and the Demographic Challenge, Teresa Ribera, the Spanish Ambassador to the United Arab Emirates, Iñigo de Palacio, and the Secretary of State for Energy, Sara Aagesen, among others. At this meeting, Teresa Ribero (2023) expressed her impression of the commitment of the Spanish business sector at the Climate Summit. Setting a precedent in the collaboration between the business sector and the Spanish government, she stressed the importance of working together to advance towards national targets for reducing greenhouse gas emissions by representing the interests of Spanish society in international forums. Economic and environmental disasters due to climate change; noting "Climate action cannot wait".

IV.- The beginning of the end of the fossil fuel era as the final outcome of COP 28 in Dubai

Countries meeting in Dubai on Wednesday approved a roadmap for the "transition away from fossil fuels," a first for a UN climate conference, but the agreement fell short of a demanded phase-out of oil, coal and gas. After the adoption of the outcome document, UN Secretary-General António Guterres said that the mention of the world's main contributor to climate change comes after many years in which the debate on this issue was blocked. Guterres (2023) stressed that the fossil fuel era must end with justice and equity.

"To those who objected to a clear reference to the phase-out of fossil fuels in the COP28 text, I want to say that the phase-out of fossil fuels is inevitable, whether they like it or not. Let's hope it doesn't come too late," he said.

The latest edition of the annual UN climate conference held in Dubai, the largest city in the United Arab Emirates. COP28 was scheduled to conclude on Tuesday, December 12, but intense late-night negotiations over whether the outcome would include a call to "phase down" or "phase out" planet-warming fossil fuels - such as oil, gas and coal - forced the conference into overtime. This major sticking point pitted activists and countries vulnerable to climate change against oil-producing nations for much of the past two weeks.

In his statement, Guterres (2023) noted that the science is clear, stating that limiting global warming to 1.5°C "will be impossible without phasing out all fossil fuels," as recognized by an increasingly broad and diverse coalition of countries. The COP28 mediators
achieved commitments to triple renewable energy capacity and double energy efficiency by 2030, and made progress on adaptation and finance,

including the launch of the Loss and Damage Fund. However, the Secretary-General considered that financial commitments are very limited and much more is needed to deliver climate justice for those on the frontlines of the crisis.

Many vulnerable countries are drowning in debt and risk drowning as sea levels rise. It is time for an increase in financing to adapt, loss and damage, and reform of the international financial architecture". Guterres (op.cit), argued that the world cannot afford "delays, indecision and half-measures", and insisted that "multilateralism remains humanity's best hope". It is essential to unite around real, practical and meaningful climate solutions that are commensurate with the scale of the climate crisis, emphasized Guterres (2023).

In that vein, UN climate chief Simon Stiell (2023) stated that at COP28, the initiatives announced in Dubai are only "a lifeline for climate action, not a victory at the finish line." Stiell (op.cit). He said the Global Stocktake, which aims to help nations align their national climate plans with the Paris Agreement, clearly revealed that progress is not fast enough, but it is "undeniably" gaining pace. Still, the current trajectory is just below three degrees of global warming, which amounts to "massive human suffering," according to the climate official, which is why COP28 should have achieved better results.

Speaking to reporters, Stiell (2023) said that COP28 should have marked a firm halt to humanity's main climate problem: "fossil fuels and their pollution, which is burning the planet". "This agreement signed, represents a set of ambitious baselines, not a complete achievement." . Thus, the coming years will be crucial to further increase ambition and action for the climate" **It should be noted that three other events occurred at COP 28, related to**

1) The Loss and Damage Fund to help developing countries vulnerable to climate change came to life on the first day of the

COP. Countries have so far pledged hundreds of millions of dollars in contributions to the fund.

2) Commitments of US$ 3.5 billion to replenish Green Climate Fund resources

3) New announcements totaling more than $150 million for the Least Developed Countries Fund and the Special Climate Change Fund.

4) An increase of $9 billion per year by the World Bank to finance climate-related projects (2024 and 2025).

5) Nearly 120 countries endorsed the COP28 Declaration on Climate and Health to accelerate action to protect people's health from growing climate impacts

6) More than 130 countries have endorsed the COP28 Declaration on Agriculture, Food and Climate to support food security while combating climate change

7) 66 countries have signed up to the global commitment to reduce refrigeration-related emissions by 68% as of today.

On the other hand, agreements were reached for the next COP meetings, such as:

- The round of national climate action plans is scheduled for 2025, when countries are expected to have seriously advanced their actions and commitments.

- Azerbaijan will officially host COP29, November 11-22 of the year (2024), after receiving the backing of Eastern European countries following Armenia's withdrawal of its bid Brazil has offered to host COP30 in the Amazon in 2025 (UNFCCC/ Kiara Worth)

However, not all delegations were satisfied with the outcome of the climate talks. Representatives of civil society and climate activists, as well as delegations from small island developing countries, were visibly unhappy with the outcome Anne Rasmussen (2023), representative of Samoa and lead mediator of the Alliance of Small Island States, noted that the decision was taken during her absence from the plenary room, as her group was still coordinating its response to the text.

Rasmussen (op.cit) underlined the importance of the Global Stocktaking process, noting that global warming can still be limited to 1.5°C." He lamented the lack of "course correction" and expressed his disappointment: "We really needed an exponential change in our actions and not business as usual. He also regretted the lack of "course correction" and expressed his disappointment: "we really needed an exponential change in our actions and not business as usual". After the publication of the outcome document, Harjeet Singh (2023), head of global policy strategy for the International Climate Action Network, told UN News that "after decades of prevarication, COP28 has finally focused on the real culprits of the climate crisis: fossil fuels.

In light of this, a course has been set to move away from coal, oil and gas. But the resolution is flawed by loopholes that offer the fossil fuel industry numerous escape routes, relying on unproven and unsafe technologies. Singh (op.cit), called the rich nations "hypocrites... as they continue to expand fossil fuel operations while talking about green transition". On the other hand, developing countries still dependent on fossil fuels are left without adequate financial support in their transition to renewables.

Although COP 28 has acknowledged the immense financial shortfall in addressing climate impacts, the final results fall short of forcing rich nations to meet their financial responsibilities. Yet $7 trillion is invested each year in activities that fuel climate change, through seagrasses, green shoot extensions and grass-like flowers, which are an effective nature-based solution to climate change. In this order, this is 30 times what is spent annually on green solutions and 7% of global GDP, according to the environment agency's report presented in Dubai.

In other words, nearly seven trillion dollars of public and private financing are spent each year on activities that directly harm nature, an amount 30 times greater than that spent annually on green solutions, according to a report presented at the Climate Change Conference (COP28) in Dubai. The text of the UN Environment Programme (**UNEP**) reveals that, despite the time taken to stop financial flows to sectors that damage humanity's most valuable assets, these investments continue, and the publication comes at a time when negotiations on the largest action ever taken in favor of climate justice are underway.

In that vein, the State of Finance for Nature report focuses on what are known as "negative financial flows for nature," underscoring the urgency of addressing the interconnected crises of climate change, biodiversity loss and land degradation. The paper highlights that these investments dwarf the annual amount invested in nature-based projects. In light of this, $5 billion of these environmentally negative financial flows come from the private sector, which is 140 times more than private investments in green solutions, with nearly half of that amount coming from five sectors: construction, electric utilities, real estate, oil and gas, and food and tobacco.

V.- To reduce emissions due to climate change

The latest report from the World Meteorological Organization (WMO) confirms that
That the last decade has been the warmest ever recorded, according to data from the
Heat trend over the last 30 years. According to its director, Petteri Taalas (2023), this is due to "greenhouse gas emissions from human activities". Marked by record land and ocean temperatures. Now, the decade of 2011-2020 witnessed a relentless increase in the concentration of greenhouse gases that turbocharged the dramatic loss of glaciers and sea level rise and the report is published as the United Nations Climate Change Conference, COP28, draws to a close in Dubai.

As the COP28 meeting draws to a close, countries agreed on a new voluntary fund to pay vulnerable nations for losses and damages suffered due to climate change. This runaway rise in temperature has had profound impacts on polar and mountain regions. In this regard, the WMO's ten-year report reveals the "drastic transformation" taking place in polar and high mountain regions, and the UN agency also warns that climate disruptions are undermining sustainable development, with dire consequences for global food security, displacement and migration.

In that vein, Taalas (op.cit), noted "We are losing the race to save our melting glaciers and ice sheets. We need to reduce greenhouse gas emissions as the planet's top and absolute priority in order to prevent climate change from spiraling out of control." On the other hand, the report paints a bleak picture, but also highlights positive developments, including the success of international efforts under the Montreal Protocol, which seeks to eliminate chemicals that deplete the ozone layer, and have reduced the hole in the Antarctic ozone layer during the period 2011-2020.

On the other hand, advances in forecasting, early warning systems and coordinated disaster management have reduced the number of casualties caused by extreme events. However, economic losses have increased; while public and private climate finance nearly doubled from 2011 to 2020, a seven-fold increase is needed by the end of the decade to meet climate targets.

VI.- Reducing refrigeration emissions a commitment of 60 countries at COP28

Refrigeration systems contribute greatly to climate change. In view of which, more than 60 countries signed the "cooling pledge" to reduce the climate impact of the refrigeration sector, which could also provide "universal access to life-saving cooling, relieve pressure on energy grids and save trillions of dollars by 2050". In light of this, the United Nations Environment Programme (UNEP) estimates that more than 1 billion people are at high risk of extreme heat due to lack of access to cooling, the vast majority in Africa and Asia.

In addition, almost a third of the world's population is exposed to deadly heat waves more than 20 days a year. Cooling provides relief to people and is also essential for other critical areas and services such as global food security and the storage and supply of vaccines. But at the same time, conventional refrigeration, such as air conditioning, is a major contributor to climate change, responsible for more than 7% of global greenhouse gas emissions. Which, if not properly managed, energy requirements for space cooling will triple by 2050, along with the associated emissions.

Generally speaking, the more we try to keep cool, the more we warm the planet. And if current growth trends continue, the consumption generated by cooling equipment, which today accounts for 20% of total electricity consumption, will double by 2050. **In that order,** current cooling systems, such as air conditioners and refrigerators, consume large amounts of energy and often use refrigerants that heat the planet. The latest UNEP report shows that taking measures to reduce the energy consumption of refrigeration equipment could lead to a reduction of at least 60% of projected sectoral emissions by 2050.

On the other hand, the cooling sector must grow to protect everyone from rising temperatures, maintain the quality and safety of food, keep vaccines stable and productive economies, said Inger Andersen (2023), executive director of the UN agency, which presented the report, said this growth should not occur at the expense of energy transition and more intense climate impacts. In view of this, the report was published in support of the *Global Cooling Commitment*, a joint initiative between the United Arab Emirates, and the *Cool Coalition* (coalition for cooling), led by UNEP.

This report describes the measures to be taken in passive cooling strategies, such as thermal insulation, ventilation, energy efficiency standards and reduction of hydrofluorocarbon refrigerants (HFCs) that heat the climate. Now, in view of that report it is recommended to follow the measures that could reduce projected emissions by 2050 of refrigeration in about 3800 million tons of CO2 equivalent. Which would mean:

- Enabling an additional 3.5 billion people to benefit from refrigerators, air conditioners or passive cooling by 2050
- Reduce end-user electricity bills by $1 trillion by 2050, and by $17 trillion cumulatively between 2022 and 2050.
- Reducing peak power demand by 1.5 to 2 terawatts (TW), almost twice the total generation capacity of the European Union today
- Avoiding investments in energy generation of 4 to 5 trillion dollars

VII.- Climate finance, in the COP28 agreements.

Gemma Duran Romero (2023), points out that the Conference of the Parties on the
The main purpose of the Climate Change Conference (COP28) held over the last few days in Dubai was to take global stock of the degree of implementation of the 2015 Paris Agreement and assess progress in limiting global warming to 1.5 °C. As in previous conferences, other issues such as fossil fuels, climate finance and the climate change loss and damage fund have also been raised. In view of is the issue of financing to address the climate change problem is key.

In relation to this, in 2023, the World Meteorological Organization published the report *United in Science*, which indicates that, between 1970 and 2021, some 12,000 catastrophes were recorded due to extreme meteorological, climatic and hydrological phenomena. In economic terms, these catastrophes cost some 4.3 trillion dollars and occurred mostly in developing countries. Now, coinciding with COP28, the report "Losses and damages, how climate change is impacting production and capital", prepared by J. Rising of Delawere University (USA), has been published.

The report includes country-by-country estimates of economic losses due to climate change. The report also indicates that extreme weather events account for losses of 1 % of GDP. These losses are 10 times higher for low-income countries and tropical regions. In view of this, in addition to the economic impact, there is a humanitarian impact due to the impoverishment of the most vulnerable population, some 3.5 billion people, according to the Intergovernmental Panel on Climate Change.

Among this population group are women and young people who are forced to migrate to other places, mortgaging the future of their communities and countries. In this regard, the climate crisis requires investments for mitigation and adaptation. This requires financing that is not always within the reach of developing countries. Moreover, this problem of investment and financing for climate crises is a long-standing one. At COP15 in 2009, developed countries committed to mobilizing some US$100 billion per year for climate action by 2020. This objective was reaffirmed at COP16. And in 2011, the Green Climate Fund was created.

The objective was to finance actions for mitigation and adaptation to climate change by the international community. However, despite the commitments made, the climate finance mobilized by developed countries is far from being achieved. In 2022, the Organization for Economic Cooperation and Development presented a balance sheet on the funds mobilized, with only $83.3 billion being provided in 2020. Of this amount, 8% went to low-income countries, which are most affected by climate change. The rest went to middle-income countries.

However, the issue of financing was taken up again at COP26, as a matter of urgency, and was included in the Glasgow Climate Pact. With it, the developed countries undertook to reach the target of US$100 billion per year by 2023. A year later, at COP27, a Loss and Damage Fund was approved to start operating on the opening day of COP28. This fund is based on a system of financial resource allocation, according to available evidence. Contributions are voluntary and come from developed and emerging countries.

The fund will be managed for four years by the World Bank. It is not specified which countries will receive the aid, but it guarantees a minimum percentage of resources for the least developed countries and small islands. This fund as such has a contribution of US$ 656 million; an amount that represents 1% of the total money committed for years, and which is far from the estimated annual costs generated by climate change, estimated at between US$ 290 and 580 billion for the year 2030.

Climate action requires access to affordable, sustainable, clean and inclusive energy, as well as adaptation measures, for which other financial resources must be mobilized. In this regard, international organizations and multilateral development banks have jointly presented concrete and urgent actions to increase financing. However, some of the most discussed innovative actions refer to the agreement between the main international financial institutions and the countries to offer climate-resilient debt clauses in their loans.

These loans will allow countries to pause debt repayment when they are affected by climate catastrophes, as announced by the United Kingdom for Senegal. The Inter-American Development Bank offered $1.2 billion in loans covered through the resilient debt clauses. In view of that, the World Bank committed to suspend debt and interest for two years in the event of a natural disaster. As a result, other institutions such as the African Development Bank, the European Bank for Reconstruction and Development and the French Development Agency have announced plans to integrate these clauses into sovereign loan agreements.

Moreover, Akinwumi Adesina (2023), President of the African Development Bank, at the Adaptation Financing Summit for Africa during COP28 on December 1, 2023, in Dubai (United Arab Emirates). Flickr / COP28 / Christopher Edralin, CC BY-NC-SA presented the creation of a hybrid financing mechanism, proposed by the African Development Bank and the Inter-American Development Bank, which will allow unused special drawing rights, in addition to reserves, to be used as lending instruments to finance climate and development. This financing channel will be through the multilateral development banks.

In view of this, other climate change-oriented investments also stand out alongside these measures. These investments were announced by the Development Bank of Latin America, the Asian Development Bank, the European Investment Bank and the Public Development Banks of the Amazon Basin Green Coalition. In terms of private investment, the Altérra

<u>fund</u>, proposed by the United Arab Emirates to improve access to climate finance for the countries of the Global South, stands out. The aim is to mobilize $250 billion worldwide by 2030.

This fund has been divided into two sections. One, Altérra Acceleration, with an endowment of $25 billion, to allocate capital directly or through investment funds. The other, Altérra Transformation, with $5 billion, for risk mitigation, helping to promote investment in the Global South. In this context, it could be concluded that COP28 has begun to put the key to the tune of financing, although not completely, because there are still countries such as the African countries that are demanding concrete actions for financing adaptation on the continent.

All this adaptation, not forgetting that funding needs to be more equitable and fair as pointed out by, Jo Adetunji (2023), Editor the Conversation.UK Academic rigor journalistic flair Moreover, one of UNEP's partners contributing to the report was Global Canopy, a data-driven non-profit organization that focuses on market factors that negatively affect nature. Its executive director, Niki Mardas (2023), told UN News that there is a group of companies or financial institutions that may be making positive investments for the environment.

They are making a lot of noise about it, but without being clear about their exposure to negative investments in nature, especially when it comes to their supply chains. Mardas (op. cit.) pointed out that, although these companies must continue to make positive investments, they must also do the hard and complex work of understanding how they are causing the problem, and they must begin to address it, involving the companies in their portfolios as well as the companies in their supply chains to change their operations and their behavior.

Mardas (op.cit) gave the example of the fight against deforestation, which is seeking to achieve a net zero emissions balance. However, only 20% of the 700 financial institutions that committed to achieving a net zero balance under the Glasgow Finance Alliance have taken any action in this

regard. In that sense, the greatest action that can be taken in favor of nature, climate and people is green finance. In view of
that, it has to be financed in a green way, but we also have to green those seven trillion dollars of financing.

Otherwise, you will always be stuck in that loop. In light of this, at a press conference in Dubai, Mirey Atallah (2023), director of UNEP's Nature for Climate Branch, said the report shows that the climate crisis continues to outpace efforts to contain it. She said funding is "the great enabler, and without money flowing in the right direction, you cannot achieve the goals that were set" at the 1992 Earth Summit in Rio to address the interconnected challenges of climate change, desertification and biodiversity loss.

Atallah (op.cit) stated that UNEP wants to use the data to show that money used to harm nature can and should be diverted to have a positive impact, and stressed that COP28 should be the tipping point. In that vein, she said that the chronic shortage of funding for nature-based solutions is not due to a lack of funds, "it's just that the money is going in the wrong direction."

To convince private companies to make the right investments, the necessary legal frameworks need to be put in place to direct funds towards nature-positive solutions. Atallah (2023) added that some private financial institutions have already started to take climate impacts into account when applying for loans, which can help to "change the direction of investments."

VIII.- ECLAC, invest between 3.7% and 4.9% of its GDP in climate finance

The Economic Commission for Latin America and the Caribbean (ECLAC) presented a **report** on the financing needs and policies required in the region for the transition to a low-carbon and climate-resilient economy, as well as current trends in regional emissions. The report highlights the importance of financing in sectors such as agriculture, livestock and forestry, which regionally account for 58% of greenhouse gas emissions. Currently, financing is mainly directed towards mitigation, to the detriment of adaptation measures.

In 2020, 89% of global climate finance was earmarked for mitigation, 8% for adaptation and only 3% for cross-cutting actions. In that order, climate change is one of the greatest challenges of our time. For years, ECLAC has analyzed its impacts in Latin America and the Caribbean and has found that the cost of inaction exceeds the cost of action (...) and that global warming will exacerbate the negative effects of extreme weather events," **warned** the Commission's executive secretary. José Manuel Salazar-Xirinachs (2023), specified that Latin America and the Caribbean set the goal of reducing emissions by between 24% and 29% by 2030, "but, to do so, the region's decarbonization rate (0.9%) would have to be four times faster".

According to the study, meeting climate action commitments also requires an investment of between 3.7% and 4.9% of regional GDP per year until 2030.

By way of comparison, in 2020 climate finance in Latin America and the Caribbean was only 0.5% of regional GDP. Therefore, closing the climate finance gap requires increasing domestic and international resource mobilization by seven to 10 times. The document specifies the investment needed for energy transition, electrification of public transport, mitigation measures to avoid deforestation, biodiversity conservation, early warning systems and poverty prevention, among other areas.

Specifically, for mitigation actions, the necessary investment would be equivalent to 2.3%-3.1% of the region's annual GDP. These funds should finance energy and transportation systems and the reduction of deforestation. The transportation sector requires the most investment. Adaptation measures require between 1.4% and 1.8% of the region's annual GDP. This includes investments in early warning systems, poverty prevention, coastal zone protection, water and sanitation services, and biodiversity protection.

In this category, the largest amounts are earmarked for water and sanitation, the paper notes. In the same vein, Salazar-Xirinachs (2023), explained that increasing climate finance can also bring other benefits in addition to environmental ones, including economic and social benefits. In this sense, more investment in mitigation and adaptation measures would be an important boost for growth, job creation and social development. Conversely, if no action is taken, climate change can lead to losses.

The report shows that, by 2030, the loss of labor productivity due to heat stress could reach 10% in some countries, which would directly affect the region's growth potential. In addition, the impact of extreme events must be taken into account. The document highlights the need to channel investment flows towards activities that stimulate the driving sectors of the economy, with a view to achieving more productive, inclusive and sustainable development. In view of this, ECLAC has identified several relevant sectors and areas of opportunity for economic growth, including the energy transition, electromobility, the circular economy, the bioeconomy, the pharmaceutical industry, digital services and the care economy, among others.

The document also specifies different instruments, such as carbon pricing and the inclusion of climate change in environmental impact assessments of projects. The document entitled "Economics of Climate Change in Latin America and the Caribbean. Financing needs and policy tools for the transition to low-carbon and climate-resilient economies (**The Economics of Climate Change in Latin America and the Caribbean, 2023. Financing needs and policy tools for the transition to low-carbon and climate-resilient economies**). This document presents current trends in regional emissions, climate action commitments and estimates of investment required to meet Nationally Determined Contributions (NDCs).

It also establishes guidelines to be followed in the quest to achieve inclusive, sustainable and fair development for the region. It was presented by the highest authority of the Economic Commission for Latin America and the Caribbean (ECLAC) during the COP28 side event "Economic cooperation between Spain and Latin America for climate finance", held in the Spanish pavilion of the world meeting, which was moderated by Gonzalo Muñoz, UN High Level Climate Champion COP25, and member of the board of GFANZ LAC.

Alicia Montalvo, Manager of Climate Action and Positive Biodiversity at CAF - Development Bank of Latin America; Ricardo Marshall, from the Roofs to Reefs Program (R2RP) of the Office of the Prime Minister of Barbados; and Elsa Velasco, Team Leader of EUROCLIMA+ 2020 at FIIAPP, also participated in the event. The paper notes that, by 2030, the loss of labor productivity due to heat stress could reach 10% in some countries, which would directly affect the region's growth potential. In addition, the impact of extreme events must be taken into account.

It highlights the importance of financing in key economic sectors such as land use change, agriculture, livestock and forestry, which account for 58% of regional greenhouse gas emissions. Currently, financing is directed towards mitigation to the detriment of adaptation and cross-cutting actions. However, according to the study, closing the climate finance gap requires increasing the mobilization of national and international resources by 7 to 10 times, said Salazar-Xirinachs (2023).

Investing in climate action can bring not only environmental, but also economic and social benefits, as the levels of investment and financing of mitigation and adaptation measures will provide a significant boost to growth, employment and social development. In the recommendations, the document highlights the need to coordinate policies and align the financial system to channel investment flows towards productive activities that drive sectors that are the engines of the economy, in order to achieve more productive, more inclusive and more sustainable development.

In this regard, he indicated that the countries of the region must intensify and scale up their productive development policies. He reiterated that ECLAC has identified several dynamic sectors, areas of opportunity for economic growth and collaboration. ECLAC remains committed and will continue to work for an environmentally sustainable, socially inclusive and economically competitive future in Latin America and the Caribbean," concluded José Manuel Salazar-Xirinachs.

IX.- Actions towards the Environmental Sustainability of the Planet

Caring for the environment and improving the quality of life is within everyone's reach. As part of **Earth Day**, *National Geographic* offers a list of activities that you can implement in your daily life to, according to the United Nations (UN), contribute to the environment and limit climate change on Earth. When we talk about <u>sustainability</u>, we refer to a development model that meets the needs of the present without compromising the ability of future generations to meet their own needs. In this sense, the following are 10 sustainable actions for the sustainability of the planet:

1. Saving light energy and taking advantage of sunlight

When you leave your home, an action to contribute to the care of resources and the environment is to make sure that the lights that are not useful are turned off. In addition, you can open the windows and let the sunlight pass through them to illuminate the house during the day.

2. Change the type of energy managed in the home.

In line with the first recommendation, replacing the lights in a home with LED bulbs is a way to contribute to the environment and save energy costs. Since they illuminate more, consume less and are amortized over a longer period.   According to the UN, a large part of our electricity runs on coal, oil and gas, which affects the emission of carbon dioxide (CO_2) into the environment.

3. Unplug unused appliances.

Save energy by reducing the use of heating, air conditioning, as well as unplugging appliances that are out of use, but still consume energy with their blinking lights on, clocks or remote control sensors.

4. Alternate transportation methods

You can alternate the use of a diesel or gasoline-powered personal <u>vehicle</u> by making short trips on foot or by bicycle, which reduces greenhouse gas emissions and improves your health performance while exercising.   The UN explains that using a car in a conscious way, replacing it even with the use of public transport for longer journeys, can reduce the carbon footprint by up to 2 tons of CO2 per year.

5. Consume more vegetables and sustainable agricultural products.

In principle, agroecological products do not use <u>fertilizers</u> or other polluting products in their production stage. In addition, UNO suggests eating more vegetables, fruits, whole grains, legumes, nuts and seeds, and less meat and dairy products, as this can considerably reduce the environmental impact.

6. Separate waste, repair and recycle

The Assembly in charge of the <u>2030 Agenda for Sustainable Development</u> warns that every element consumed by human beings generates carbon emissions in each link of the production chain (household appliances, clothing, sundries). This is why the UN suggests mending clothes that are still useful, making purchases that are conscious of personal needs and recycling what is no longer used.   In addition, people can recycle garbage by separating waste into organic (food), non-organic (paper) and plastic.

7. Use less plastic

One way to use less <u>plastic</u> is to carry cloth bags, burlap or reuse plastic bags when shopping. More and more supermarkets are selling bags to avoid their use and generate a surplus cost to the customer and encourage recycling, says UNO.

8.- Waste less food

When you throw food away, you also waste the resources and energy that went into growing, producing, packaging and transporting it," says the international agency. Decomposing food, like animal waste, produces a greenhouse gas called <u>methane</u>.

9. Composting organic waste

Reusing food and animal waste in the production of compost is an effective method that UNO recommends to waste less food and <u>compost</u> a household's plants naturally. Reducing food waste can reduce the carbon footprint by up to 300 kilograms of CO2 per year.

10. Tree planting and seeding

Plants are a natural source of life and produce the oxygen that living things breathe on Earth. They are essential to nature, therefore, UNO recommends planting a tree or shrubs in your home and/or in the community where you live.

On the other hand, in order to contribute to sustainability it is necessary to understand that the problems affecting sustainability are not restricted to large companies, since in one way or another we all contribute to the sustainability of the planet. In that order, the solutions to the problems that affect sustainable development should not be limited only to policies, strategies and standards designed and established in companies. Although they may seem insignificant, our individual actions can contribute considerably and positively to sustainability, it is the awareness to achieve a truly sustainable development. The following is a set of measures that should be considered to contribute to this very important cause:

Reduce (do not waste resources)

- Control water consumption in hygiene, irrigation and swimming pools.
- Incorporate water saving devices in faucets and cisterns.
- Quick shower; turn off faucets while brushing teeth, shaving or soaping up.
- Proceed to drip irrigation, watering early and late in the day.

- Reduce energy consumption in lighting, use energy-saving light bulbs: compact fluorescent and LED (Light Emitting Diode).
- Turn off unnecessary lights (to overcome inertia) and make the most of natural light.
- Use motion sensors to turn on the light only when it is needed
- Reduce energy consumption in heating, cooling and cooking.
- Insulate (apply appropriate insulation standards to dwellings)
- Do not program temperatures that are too high (dress more warmly) or too low (ventilate better, use awnings, blinds...); use a timer and place thermostats in suitable locations.
- Turn off unnecessary radiators or air conditioners (overcome inertias)
- Cook efficiently: use residual heat, do not heat more water than necessary and do not preheat in the oven if it is not necessary.
- Reduce energy consumption in transportation, use public transportation, cycling and/or walking.
- To organize the displacements of several people in the same vehicle.
- Reduce speed, drive efficiently.
- Avoid elevators whenever possible.
- Properly load washing machines, dishwashers, etc.
- Completely turn off the TV, computer and other electrical appliances.
- When not in use; unplug chargers for cell phones and other electronic devices.
- Reduce battery consumption or use rechargeable batteries.
- Regularly defrost the refrigerator, check that the doors close properly, check boilers and heaters.
- Reduce energy consumption in power supply, while improving it at the same time.

- Consume seasonal and organic products.
- Reduce the use of paper, avoid printing documents that can be read on the screen.
- Writing, photocopying and printing double-sided and taking advantage of space (without leaving excessive margins).
- Avoid commercial mail; delete yourself from the databases of advertising companies.
- Congratulate, communicate and convene meetings electronically.
- Use recycled paper.
- Reduce (better to avoid!) the use of plastics, cans, objects with batteries, materials with toxic substances, etc.
- Reduce the consumption of plastics, particularly PVC, in toys, footwear, small household appliances, cleaning products and others (if unavoidable), choose recyclable materials (PET, HDPE, etc.), reusing them as much as possible.
- Avoid battery-operated electrical appliances and toys.
  Reduce the consumption of products containing toxic substances, such as insecticides, solvents, disinfectants, stain removers, polishes, aggressive cleaning products ("cleaning without chlorine"), do not buy clothes that must be cleaned in dry cleaners or use ecological dry cleaners, etc.
- Reject consumerism: practice and promote responsible consumption.
- Critically analyze advertisements (see www.consumehastamorir.com).
- Do not get carried away by commercial campaigns: Valentine's Day, Epiphany, among others.
- Schedule shopping (go shopping with list of needs)

Reuse as much as possible

- Reuse paper

- Printing, for example, on paper already used on one side
- Reuse water: use water from washing fruits and vegetables and from cooking eggs (enriched with calcium) to irrigate plants.
- In particular, avoid plastic bags and wrapping, aluminum foil and paper cups.
- Replace them with reusable ones, repairing them when necessary, using recycled products (paper, toner, etc.) and recyclable products as long as it is possible.
- Rehabilitation of dwellings, making them more sustainable (better insulation, etc.) and avoiding new construction.)

Recycle

- Separate waste for selective collection ("compacting" it so that it takes up less space).
- Take to "Puntos Limpios" what cannot go to ordinary dumps.
- Recycle batteries, cell phones, light bulbs containing mercury, computers, oil, toxic products.

Use of environmentally and people friendly technologies

- Do not buy products without making sure they are safe: check the composition of food, cleaning products, clothing, etc., and avoid those that do not offer guarantees.
- Avoid sprays and aerosols (use manual sprayers).
- Apply safety rules at work and at home.
- Opting for renewable energies in the home, automotive, etc.
- Use solar-powered appliances: radios, cell phone chargers, laptops.
- Use efficient, energy-efficient and low-pollution appliances (A++).

Contribute to citizen education and action

- Carry out outreach and promotion tasks: take advantage of the press, Internet, video, ecological fairs and school materials.

- To help raise awareness of unsustainable and closely linked problems: consumerism, population explosion, predatory economic growth, environmental degradation and imbalances.
- To inform about the actions that we can carry out and encourage their implementation, promoting campaigns for the use of low-energy light bulbs, reforestation and associations.
- To help conceive sustainability measures as an improvement that guarantees the future of all and not as a limitation.
- Promote social recognition of positive measures for a sustainable future.
- Study and apply what you can do for sustainability as a professional.
- Research, innovate and teach.
- Contribute to the environmentalization of the workplace, the neighborhood and the city where we live.

Participate in socio-political actions for sustainability.

- Respect and enforce environmental protection legislation for the protection of biodiversity.
- Avoid contributing to noise, light or visual pollution.
- To express to businesses our disagreement with the use of excessive wrapping, waste of plastic bags, non-separation of garbage, etc.
- Do not smoke where third parties are harmed and never throw cigarette butts on the ground.
- Do not leave waste in the forest, on the beach.
- Avoid residing in developments that contribute to the destruction of ecosystems and/or higher energy consumption.
- Be careful not to damage the flora and fauna.
- Comply with traffic regulations for the protection of people and the environment.

- Denounce the policies of continued growth, which are incompatible with sustainability.
- Denounce ecological crimes: illegal logging, forest fires, untreated waste, predatory urban planning.
- Respect and enforce respect for human rights, denounce any ethnic, social and gender discrimination.
- Collaborate actively and/or financially with associations that defend sustainability.
- Support programs to help the Third World, defend the environment, help populations in difficulty and promote Human Rights.
- Claiming the application of solidarity taxes
- Promote Fair Trade.
- Reject products resulting from predatory practices (tropical timber, animal skins, fishing, etc.).
- The company's activities are not only a source of income (e.g., for the exploitation of the environment, unsustainable tourism...) or are obtained with labor without labor rights, child labor and support companies with a guarantee.
- Demand clear information policies on all issues.
- Defend the right to research without ideological censorship.
- Demanding the application of the precautionary principle
- Opposing unilateralism, wars and predatory policies
- Demand respect for international legality.
- Promote the democratization of world institutions (IMF, WTO, WB, etc.).
- Respect and defend cultural diversity
- Respect and defend the diversity of languages.
- Respect and defend knowledge, customs and traditions (as long as they do not violate human rights).
- Vote for parties with policies more favorable to sustainability.

- Working to get governments and political parties to take up the defense of sustainability
- Demand local, state and universal legislation to protect the environment "Cyberact": Support solidarity and sustainability campaigns from the computer.

Evaluate and compensate

- Conduct personal behavioral audits
- Adequately track our contributions to sustainability in housing, transportation, professional and citizen action.
- Compensate for the negative repercussions of our actions when we cannot avoid them (emissions, emissions of greenhouse gases, etc.)
- emissions, use of polluting products...) through positive actions (See www.ceroco2.org in effective momentum, favoring positive results and stimulating a growing involvement.
- Initially, it is advisable to select those measures that are most feasible and to agree on plans and forms of follow-up that will provide effective impetus, promote positive results and stimulate growing involvement.
- NOTE: These measures were taken from the working document produced as a contribution to the Decade of Education for a Sustainable Future (2005-2014) instituted by the United Nations to address the current planetary emergency (www.oei.es/decada).

X.- Sustainable Innovations

A greater push for sustainable products and technologies like the one we have today. A critical point has been reached with respect to climate change and many innovators and companies are stepping forward to build a greener future. In this context, as the energy transition for the drive to sustainable economy progresses, the first sustainable innovations 2023 explore energy storage challenges, Artificial Intelligence projects to address food waste, among others, are also included. The following are some of the advances that seek a more sustainable way of living:

- Iron battery to store energy at the grid level.

While lithium-ion batteries have become ubiquitous in products such as electronics, small and large appliances, electric vehicles and electrical energy storage systems, there are serious problems associated with them containing numerous toxic metals that make their manufacture, recycling and use problematic for the environment. Startup Form Energy, which came out of the Massachusetts Institute of Technology (MIT), has found a way to make metal batteries more efficient.

Although zinc batteries currently used in hearing aids contain less toxic materials, because they are not rechargeable, they are also part of the waste. However, they have found a way to reverse the corrosion process to create rechargeable iron batteries which are heavier than lithium-ion batteries and have a slow charging and recharging cycle, making them unsuitable for use in electric vehicles. However, the company claims that they will be perfect for grid-level energy storage, as they excel in long-term energy storage and can generate more than three megawatts in production capacity per battery acre.

- Subway hydrogen storage with gravity.

It is increasingly common for green hydrogen, the cleaner, zero-carbon variety, to be touted as a crucial element in the world's journey to net zero. But storing the clean fuel remains a challenge. That is why

Gravitricity, a UK-based subway energy storage specialist, is completing its design of specially engineered subway rock shaft liners, which would enable efficient subway storage of hydrogen.

Moreover, its storage technology, which it calls FlexiStore, is a solution to the obstacles facing hydrogen storage, providing a much larger and safer system; it is more flexible than subway salt caverns, another method for storage. On the other hand, Gravitricity has identified many sites for its pilot project in the UK, where discussions and future commercial plans are underway.

In 2024, companies must not only respond to rising social and regulatory expectations, but also position themselves as proactive leaders in creating a sustainable and ethical future. The combination of stricter regulations, transparent accountability and active collaboration with the public and financial sectors will lead the way to positive and lasting business transformation. Cristina Sanchez (2023) CEO of the UN Global Compact Spain. This article addresses some of the trends from Euromonitor International and some of IMD's experts, showing the most relevant issues for 2024 and how they relate to the topic of waste.

- Moving towards standardization: Sustainability reporting

One of the most notorious trends is the advance towards the generation of sustainability reports so that companies can communicate the actions they are taking regarding sustainability. For which new requirements have arisen both internationally and in the European Union (EU), which have an impact on Mexico. At the international level, on January 1, 2024, two new standards of the International Sustainability Standards Board (ISSB) came into force, which will be adopted in Mexico once the National Banking and Securities Commission establishes whether they are applicable to the entities under its regulation.

- ✓ IFRS S1 "General Requirements for Disclosure of Financial Information Related to Sustainability".
- ✓ IFRS S2 "Climate-related disclosures".

Also, in the EU, the Corporate Sustainability Reporting Directive (CSRD) came into force on January 5, 2023, requiring EU companies to report on the impact of their business activities on the environment and society and on the impact of their <u>environmental, social and governance (ESG) initiatives</u> on the business side. With the entry into force of these standards, there will be new challenges for multinational companies to comply with the various requirements and to align reporting methodologies.

The reporting landscape will allow investors, analysts, consumers and others to assess sustainability performance and identify and mitigate potential environmental risks (such as waste), as well as consolidate information for better decision making, allowing companies to increase their resilience.

- Development of circular business models

As a result of regulatory incentives such as the European Commission's Circular Economy Action Plan, the momentum towards the circular economy has been growing. But this trend is not only in Europe: in Mexico, it can be seen in the General Law of Circular Economy and in the National Strategy for Circular Economy. More and more companies have developed profitable business models based on the circular economy, thus creating changes in business dynamics (some examples are Grupo AlEn, Arca Continental, Bio Pappel, among others). From the issue of waste management in the recycling industry and recoverable waste, allowing to double revenues with circular products and services.

- Transparency in supply chains

The Corporate Climate Responsibility Monitor report, which analyzes the climate change plans of 24 multinational companies, "The climate strategies of most companies are mired in ambiguous commitments, compensation plans that lack credibility, demand for greater transparency on corporate issues from investors means that companies need to take more rigorous measures to be clear about the processes they follow and the actions they take. Good traceability of the supply chain from suppliers, manufacturing, distribution and the end customer is required to provide consumers with clear, credible and easily accessible information. The main approaches to be considered for this trend are:

a) Companies must substantiate their environmental claims using standard methodologies.

b) Obtaining third-party certifications and endorsements.

c) Do not present unproven durability data in terms of time or intensity under normal conditions of use.

In the case of Mexico, this trend becomes important due to the publication of the first edition of the Sustainable Taxonomy of Mexico in March 2023, which presents a classification system to identify and define activities, assets or investment projects with positive impacts on the environment and society, contemplating predefined goals and criteria. The taxonomy will be a key point to mitigate the risk of greenwashing, while at the same time allowing investments in sustainable activities to be encouraged, thanks to greater transparency on the part of companies.

- Sustainability as a driver of profitability and strategy

Sustainability is beginning to be seen as a key aspect to drive cost efficiency. Companies are being encouraged to align their sustainability goals with their financial objectives. According to 2023 data from Euromonitor Voice of the Industry, companies have invested or plan to invest heavily in recycling as a way to reduce their costs and reduce their environmental impact. A critical point has been reached with respect to climate change and many innovators and companies are stepping forward to build a greener future.

In this context, as the energy transition for the drive towards a sustainable economy progresses, the first sustainable innovations (2023) explore energy storage challenges, Artificial Intelligence projects to address food waste, among others, are also included. The following are some advances that seek a more sustainable way of living:

To achieve this, it has developed a process that sterilizes food waste and converts it into an organic material that can absorb up to 30 times its mass in water. This material can be applied to crops, decreasing the need for chemical fertilizer and also ensuring more efficient water distribution. Trials in the Middle East have resulted in yield increases of up to 40% and a reduction in water requirements of up to 50%. As a result, Aquagrain was one of four winners of the Foodtech Challenge.

The winners were announced at Abu Dhabi Sustainability Week and shared a $2 million grant funding award.

Artificial intelligence to address food waste in restaurants. In many developed nations, such as the United Kingdom, the majority of food waste produced nationally each year occurs at the point of customer use. Restaurants, cafes and grocery stores produce more waste than is generated throughout the supply chain In this regard, Orbisk has come up with a digital solution to this challenge, providing cameras and artificial intelligence software to professional kitchens.

Such cameras, after observing food waste patterns, can quantify and predict food waste trends. This can help kitchen staff not to over-order or over-prepare food or to modify menus to reduce the size of meal items that are often wasted. Investing in sustainability strategies can also help attract and retain talent, engage with stakeholders, meet consumer demands and drive innovation.

- Carbon accounting

Businesses should quantify the greenhouse gas (GHG) emissions they produce directly or indirectly from their activities to gain a better understanding of their environmental impact. They can then set clear emission reduction targets that demonstrate their commitment to decarbonization. In October 2023, the European Union's Carbon Border Adjustment Mechanism (CBAM) entered the transition phase, requiring importers to report the emissions associated with their products by January 31, 2024.

However, in order to promote best practices, greater transparency on corporate issues by investors means that companies have to take more rigorous measures to be clear about the processes they follow and the actions they take. Good traceability of the supply chain from suppliers, manufacturing, distribution and the end customer is required to provide consumers with clear, credible and easily accessible information.   The main approaches to be considered for this trend are:

a) Companies must substantiate their environmental claims using standard methodologies.

b) Obtaining third-party certifications and endorsements.

c) Do not present unproven durability data in terms of time or intensity under normal conditions of use.

In the case of Mexico, this trend becomes important due to the publication of the first edition of the Sustainable Taxonomy of Mexico in March 2023, which presents a classification system to identify and define activities, assets or investment projects with positive impacts on the environment and society, contemplating predefined goals and criteria. The taxonomy will be a key point to mitigate the risk of greenwashing, while at the same time allowing investments in sustainable activities to be encouraged, thanks to greater transparency on the part of companies.

- Sustainability as a driver of profitability and strategy

Sustainability is beginning to be seen as a key aspect to drive cost efficiency. Companies are being encouraged to align their sustainability goals with their financial objectives. According to Euromonitor Voice of the Industry 2023 data, companies have invested or plan to invest heavily in recycling as a measure to reduce their sustainable innovations. There has never been a greater push for sustainable products and technologies than there is today.

A critical point has been reached with respect to climate change and many innovators and companies are stepping forward to build a greener future. In this context, as the energy transition for the drive towards a sustainable economy progresses, the first sustainable innovations 2023 explore energy storage challenges, also included are Artificial Intelligence projects to address food waste, among others. The following are some of the advances that seek a more sustainable way of living:

- By developing a process that sterilizes food waste and converts it into an organic material that can absorb up to 30 times its mass in water.
- This material can be applied to crops, decreasing the need for chemical fertilizer and also ensuring more efficient water distribution.
- Tests in the Middle East have resulted in yield increases of up to 40% and a reduction in water requirements of up to 50%. As a result, Aquagrain was one of the four winners of the Foodtech Challenge.
- The winners were announced at Abu Dhabi Sustainability Week and will share a $2 million grant funding award.
- Artificial intelligence to address food waste in restaurants. In many developed nations, such as the United Kingdom, the majority of food waste produced nationally each year occurs at the point of customer use.
- Restaurants, cafeterias and grocery stores produce more waste than that generated along the supplychain
- In this regard, Orbisk has come up with a digital solution to this challenge, providing cameras and artificial intelligence software to professional kitchens.
- These cameras, after observing food waste patterns, can quantify and predict food waste trends.
- This can help kitchen staff not to over-order or over-prepare food or modify menus to reduce the size of meal items that often go to waste. Costs.

- Investing in sustainability strategies can also help attract and retain talent, engage with stakeholders, meet consumer demands and drive innovation.

Because of this adjustment, the price of imported goods will depend on the emissions emitted in their processes, for which Mexico must prepare itself, since this carbon price may affect products going to the EU. 92% of the world's gross domestic product (GDP) has declared its intention to commit to achieving zero net emissions by 2050 (Net Zero Tracker, 2023). Overall, the new landscape for 2024 shows a strong trend towards changes in the way companies conduct their operations, allowing for greater transparency in sustainability decision-making, generating commitments to reduce emissions, and generating reports to communicate progress.

Now, taking actions within these axes can contribute to acquire a better positioning as a company, prepare for possible environmental and social risks, have a better positioning within the financial environment and in general, allow the company to have greater resilience. This year, 2024, is aimed at creating conscious changes to contribute to global objectives and comply with regulations, it is time to act accordingly and be part of this sustainable transformation as a company.

XI.- Raising awareness and reflection

In 1992, March 26 was declared World Climate Day by the United Nations Framework Convention on Climate Change. This day was created to raise awareness and sensitize about the importance and influence of climate and the impact of climate change on human beings worldwide. Days like this or the recent International Day of Forests, are an invitation for citizens to reflect on lifestyles and the way we behave with respect to the environment. And, of course, to be aware that our actions not only affect us human beings but also the rest of living beings and the natural resources that exist.

World Climate Day is also related to one of the Sustainable Development Goals (SDGs), specifically SDG 13 on climate action, which aims to introduce climate change as a key issue in the policies, strategies and plans of countries, companies and civil society, improving the response to the problems it generates, and promoting education and awareness of the entire population in relation to the phenomenon. However, how can we increase the environmental awareness of society, in that sense we are aware of the importance of taking care of our <u>health</u> and we must also be aware of the need to protect our environment.

Environmental awareness is a necessary learning process, regardless of our age or **our** knowledge. In this order, environmental awareness is the basis of everything, besides being a philosophy of life that cares about the <u>environment</u> and protects it in order to preserve it and ensure its present and future balance. In view of this. We must be aware that one of the aspects that deteriorate nature the most is man. Deforestation, air pollution, water pollution and global warming are consequences of the lifestyle that prevails in our society.

Likewise, environmental education and environmental awareness help us realize that every action we take in our daily lives has an impact on the environment. The means of transportation we use to go to work, the use of plastic bags, the type of energy we consume, all have an influence. In this sense, the awakening of environmental awareness is done through education and awareness, and because environmental awareness can be promoted in two ways:

- From school, through environmental education exercises for children.
- Through initiatives to raise awareness of the consequences that our actions can have on the environment.

At school, practices can be carried out such as sorting solid waste to throw each item in its corresponding container; activities focused on the reuse of materials, and visits to natural parks to observe animals in their natural habitat, which helps to understand why it is essential to protect natural resources. These types of activities awaken environmental awareness from childhood and give rise to generations that are more respectful of nature and their environment.

On the other hand, awareness-raising actions to promote environmental awareness can be very diverse: from specific events on specific topics to advertising campaigns that make us reflect on our daily habits and how they affect nature. In short, environmental education and environmental awareness invite us to change our daily habits and to open our eyes to see what is going on around us.

XII.- Conclusions

- SEO/Birdlife's COP28 (2023), together with other science and civil society movements, will ensure climate commitments to integrate conservation and restoration of the natural environment for the coming years.
- Sustainability is key to driving cost efficiency. Looking for companies to match their sustainability goals with their financial objectives.
- Slowing the advance of climate change through warnings of the impacts of climate change on nature and the need to act quickly and effectively to halt it and maximize resilience.
- Halt the development of fossil fuel production, incompatible with its human rights obligations and the goal of limiting global warming to below 1.5°C.
- triple the capacity of renewable energy by 2030, while respecting biodiversity and local communities in its facilities
- low technological incorporation, and the educational systems have not been able to implement quality alternative teaching methods that would allow for the continuation of service provision in the face of climatic emergencies.
- Fight against climate change by generating contributions from schools for environmental sustainability, accompanying them and adding value to the countries' agenda of decarbonization and resilience to climate change.
- The actions of companies have to go towards mitigation, climate finance, adaptation and restoration of biodiversity.
- Strengthen international cooperation in terms of incentives, regulations and conditions to guide investments to achieve a global transition towards the reduction of CO2 emissions.

- It is time to increase funding for adaptation, loss and damage, and reform of the international financial architecture, with "multilateralism being humanity's best hope".

- **According to the Environment agency the investment of seven trillion dollars for activities such** as seagrass meadows, expanses of green shoots and grass-like flowers, which are a nature-based solution, accounts for: 30 times what is spent annually on green solutions and 7% of global GDP.

- UNEP will adopt measures to reduce the energy consumption of refrigeration equipment to reduce at least 60% of the sectoral emissions expected by 2050.

- UNEP's **Global Cooling Commitment** outlines passive cooling strategies to reduce cooling emissions by 3.8 billion tons of CO2 equivalent by 2050.

- **International organizations, multilateral development banks and countries presented a set of urgent and innovative concrete actions to increase financing through climate-resilient debt clauses in their lending.**

- **The World Bank committed to suspend debt and interest for two years in the event of a natural disaster.**

- **The European Bank for Reconstruction and other institutions announced plans to integrate these clauses into sovereign loan agreements.**

- The chronic shortage of funding for nature-based solutions is not due to a lack of funds, "it's just that the money is going in the wrong direction.", the necessary legal frameworks need to be put in place to direct funds towards nature-positive solutions.

- **ECLAC,** points out the financing needs for the transition to a low-carbon and climate-resilient economy, as well as the current trends in regional emissions.

- Finance sectors such as agriculture, livestock and forestry, which account for 58% of greenhouse gas emissions. This funding is aimed at mitigation, to the detriment of adaptation measures.
- The need to channel investment flows toward activities that stimulate the economy's driving sectors, with a view to achieving more productive, inclusive and sustainable development.
- Implement activities *in your daily life that* contribute to the environment and limit climate change on Earth. Our individual actions can contribute significantly and positively to sustainability, to achieve a truly sustainable development.
- To increase society's environmental awareness, to protect our environment, as well as to lend importance to our health.
- Environmental awareness is a necessary learning process, regardless of our age or knowledge.
- Environmental awareness helps us to realize that every action we take in our daily lives has an impact on the environment.
- The awakening of environmental awareness is done through the school with environmental education and awareness of the consequences of our actions on the environment.
- Raise individual awareness to change habits in order to stop the irreparable effects of climate change on the life of the planet.

XIII.- References

Jo Adetunji (2023) **funding must be more equitable and fair** Editor the Conversation.UK Academic rigor journalistic flair **https://blog.theconversation.com-uk,**

Adesina Akinwumi (2023),**creation of a** hybrid **financing mechanism, proposed by the African Development Bank and the Inter-American Development Bank** Https://**www.afdb.org**

Clara Arpa (2023) **Climate Ambition Accelerator** program https://www.goaragon

Inger Andersen (2023), **this growth must not come at the cost of energy transition and more intense climate impacts** Https://www.unep.org-people

Mirey Atallah (2023), **said the report shows that the climate crisis continues to outpace efforts to contain it**. Director of UNEP's Nature for Climate Branch, Https:// unep.org- World-ci

Gemma Duran Romero (2023) **the issue of financing to address the problem of climate change is key.** https:// portalcientifico.uam.es

EuroNews (2023), **COPS28 a historic agreement to abandon fossil fuels** https://es.euronews.com

El País (2023) **environmental, economic and social sustainability** https:// elpaís.com-environment and sustainability

Aquae Foundation (2023) **Sustainable actions to take care of the planet https://www.fundacionaquae.org** ' acciones-sostenibles

Antonio Guterres (2023) **Limiting global warming to 1.5°C, will be impossible without phasing out all fossil fuels** Https:// www.un.org.Guterres

IPPC (2023) **New analysis on national climate plans** Https://unfccc.int-news-new-analysis on climate plans

Liora Schwartz, Maria Soledad (2023), **Education and Climate Change, how to develop skills, for school age climate action** IDB. Https://blogs.iadb.org-educaction,author

Niki Mardas (2023) **Is speaking at. Business models that enrich biodiversity: How can we embed nature at the heart of business?** Https://globalcanopy.org-news

OCHCR.org (2023) OHCHR and Climate Change Https://.wwwochcr.org-climate Change

Ojiambo Sandra (2023) **Executive Management Team One Global Compact** Https://unglobalcompact.org

UN (2023) **Action** on **climate change cannot waitCOPS28, Dubai** **https://.www.un.org-** **climate Change**

Anna Rasmussen, (2023) the importance of the Global Stocktaking process, noting that global warming can still be limited to 1.5°C". https:// climática.coop- interview .anne rasmussen

Teresa Ribeiro (2023) UN **climate change Pavilion at Cop 28** https:// www.unfccc.int -cop28

Salazar- Xirinachs (2023), **increasing climate finance can also bring other benefits in addition to environmental ones, including economic and social benefits** Https://www.cepal.org-equipo-jose manuel Salazar xirinachs

Cristina Sánchez (2023) **The need to transform the global financial system.** UN Global Compact Ceo in Spain https://pactomundial.org-noticias

Stiell Simon (2023) **the world walks too slowly in the face of a terrifying climate crisis https://reliefweb.int-report,** World-cop28

Harjeet Singh (2023) **getting rich countries to pay up for climate** https://axios.com harjjeet

Petteri Taalas (2023) **ends 8 years of fight against climate change** https: //www. Infobae.com -agencies

I want morebooks!

Buy your books fast and straightforward online - at one of world's fastest growing online book stores! Environmentally sound due to Print-on-Demand technologies.

Buy your books online at
www.morebooks.shop

Kaufen Sie Ihre Bücher schnell und unkompliziert online – auf einer der am schnellsten wachsenden Buchhandelsplattformen weltweit! Dank Print-On-Demand umwelt- und ressourcenschonend produziert.

Bücher schneller online kaufen
www.morebooks.shop

Printed by Books on Demand GmbH, Norderstedt / Germany